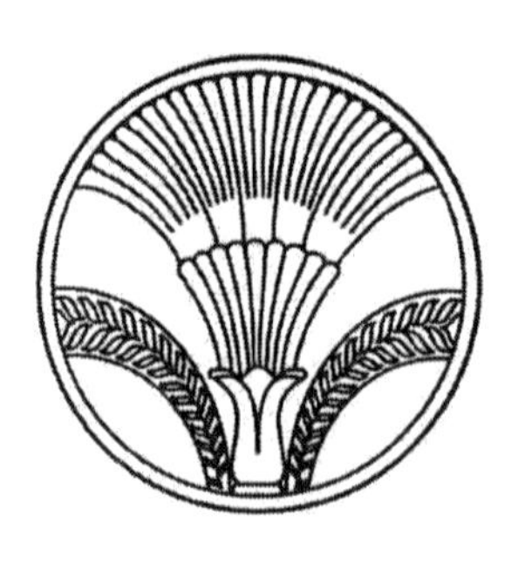

DEPARTMENT OF AGRICULTURE, VICTORIA,

ILLUSTRATED DESCRIPTION OF THISTLES, ETC.,

INCLUDED WITHIN THE PROVISIONS OF THE

THISTLE ACT OF 1890.

By Authority:
ROBT. S. BRAIN, GOVERNMENT PRINTER, MELBOURNE

1893.

2921.

DEPARTMENT OF AGRICULTURE, VICTORIA.

ILLUSTRATED

DESCRIPTION OF THISTLES,

ETC.,

INCLUDED WITHIN THE PROVISIONS OF THE

THISTLE ACT OF 1890.

By Authority:

ROBT. S. BRAIN, GOVERNMENT PRINTER, MELBOURNE.

1893.

2921.

DESCRIPTIONS AND ILLUSTRATIONS

OF

THISTLES AND ALLIED PLANTS

REQUIRED TO BE DESTROYED UNDER THE THISTLE ACT OF 1890.

THIS unpretensive publication arose from a desire of facilitating an exact knowledge of those plants which, under the above-mentioned Statute, come to public notice, and should therefore be elucidated in a manner admitting of no uncertainty of recognition. Chromo-lithographic illustrations, as well as descriptions of all the Thistles and thistle-like plants mentioned in the Act are now provided, so that anywhere the various species might be identified locally with more ease, whereby the necessity for obtaining departmental information in this direction will be very much lessened. Further on in this introductory text the principal characteristics of the Thistles here under consideration have been briefly contrasted; thus as far as possible tedious comparisons of the lengthened descriptive records will be avoided, should glances on the respective pictures not prove at once sufficient for any information anywhere sought.

As for rural purposes it is desirable that the naming of the Thistles here hitherto immigrated and copiously spread should be as free from complication as possible, it was deemed best to adhere to the systematically scientific names, the vernaculars being subject to much local vacillation; but to simplify the nomenclature, the generic name Carduus in the Linnéan sense is maintained, so that the modern generic appellations Cnicus, Cirsium and Silybum are given merely as synonyms for the true Thistles. Perplexing popular names of very limited significance are omitted. The descriptions may seem unnecessarily effuse; but to distinguish the Thistles, now already here required to be elucidated, from allied additional species, which are sure at some future time to invade our colonial territory, a full account of each sort becomes needful. In identifying any kind of true Thistles, it should also be borne in mind that hybrids occur not rarely among them, rendering their specific recognition not always easy. It has further been considered desirable to offer some few remarks on the best method of suppression of these troublesome weeds, not for experienced farmers, but for such settlers, who as colonists enter on rural pursuits for a new occupation of life. In coping with plants of this kind two distinct means must be kept in view as necessary. When annual weeds are to be dealt with, the difficulty is not so great as in those cases where perennial plants have to be subdued. Whatever method is adopted for getting rid of plants seeding only once from the same root when left undisturbed, it reduces itself to acting on the principle of destroying the plants before they can ripen any seeds. In the case of Thistles and allied plants of annual growth, when the individual plants are cut or lifted, they should immediately be brought into heaps and be burned at once, otherwise the seeds may ripen still on the plants; inasmuch as though dislodged or cut, they are apt to retain vitality for some considerable time while on the ground, all Thistles being succulent and slow in drying up. The difficulty of dealing successfully with Thistles and other weeds of perennial growth is much greater; and regretably the worst of all European Thistles, Carduus arvensis, has in late years made its appearance also here. The difficulty of its eradication is augmented, because the root is ramified and fragile; besides, the smallest portion left near the surface will continue to grow. Here it may incidentally be mentioned, what Dominique Villars already asserted, in the last century, that Carduus arvensis could best be subdued by not interfering with it till after flowering. So it would seem, that the root gets much exhausted by the far advancing growth of the plant, without the danger of seeds ripening being readily incurred by this almost unisexual Thistle. Villars pointedly says, that by any early cutting the plant would increase in all directions from the root. As this Carduus is in our colony of comparatively recent introduction, this advice could not yet be much brought to test here; it is not applicable however to Thistles of only annual duration, and would probably be neither of avail, when Carduus arvensis has become a plant of several years' age.

It would seem, that for providing an expeditious, effective, and uncomplicated "Weeds-Destroyer," the ingenuity of invention can still further be exercised, with a view of superseding to some extent manual application by usual hoes, forks, scythes, spades, ploughs or other tools.* Beyond the ordinary operations and implements in rural use for the destruction of weeds of various kinds, some contrivances exist perhaps not yet adopted here. Thus attention is directed, where locality admits of application, to a particular "Paring Plough," which cuts the roots readily under the surface, and thus prevents the plant forming several side-stems, as it is prone to do when cut just above ground. In some cases depasturing can be resorted to, particularly by sheep, and perhaps goats and pigs, as the continued deprivation of stem and leaves becomes at last detrimental to the root; thus sheep will nip off the young sprouts of Carduus arvensis with avidity; and if this is carried on sufficiently long, the roots will perish of exhaustion. The

* Here the "Thistle-hoe" (a small lifting spade) and the "Thistle-chopper" are chiefly utilized.

destruction of the small Rumex-Sorrel (Rumex Acetosella, L.) can be aided by sowing broadcast repeatedly turnips between, and feeding off through sheep the united young growth, the sorrel as the weaker plant getting suffocated and tramped out. Pigs will devour as well Thistle-roots as Bracken-roots, and by keeping the animals within hurdles, moved gradually over the ground, the annihilation of these plants can be effected more readily on some places than by other means. It should be further taken into consideration, that in the winterless clime of our lowlands the growth of weeds proceeds more or less through the whole year, and this renders coping with such plants here far more onerous than in countries where the length and severity of the winters annihilate largely such plants and their seeds. Therefore, the Australian ruralist has far more perseveringly to bring his operations to bear for the suppression even of mere annuals. To deal with obtrusive plants of perennial growth is, of course, still far more complicated, as our experiences show with such plants, for instance, as the Sorrel-Rumex, the Bracken-Fern, the Furze, the Briar-Rose. Hitherto the Thistle and cognate plants which have invaded our colony are all annuals, except Carduus arvensis, above alluded to. This can further be distinguished from the other Thistles, with us hitherto naturalized, by perfect male and perfect female flowers being developed by separate individual plants only. Carduus arvensis is a "Plume-Thistle," that is to say, the tuft on the summit of the seed-like fruits consists of plumous bristlets. As yet only one more real Plume Thistle is widely naturalized in our colony, the very common Carduus lanceolatus, the other remaining genuine Carduus, namely C. pycnocephalus, has simply hair-like bristlets of the tuft. These easy distinctions as regards the "pappus" are observable already in the young flower-headlets.

For further facilitating the use of the plates and the descriptions of the nine plants recorded here, it may introductorily be noted that the Heraldic Scotch or Onopordon-Thistle, which in its less grey state bears some resemblance to Carduus lanceolatus, has the inside bottom of the flower-headlets, so to say, honeycombed, and not beset with capillulary bristlets; while the Spotted Thistle, Carduus Marianus, is known well enough everywhere by the vernacular name, signifying as if a milky fluid had been sprinkled on the foliage. The Centaureas are spurious Thistles; and as we have hitherto to attend only to two, their discrimination is quite easy by the mere colour of the flowers, purple in Centaurea Calcitrapa, yellow in Centaurea Melitensis. The only other yellow-flowering plant coming under notice on this occasion is the very conspicuous Saffron-Thistle, Kentrophyllum lanatum. Thus it solely remains under the Thistle Act to recognise the so-called Bathurst-Burr, a plant very different from the eight others treated in this essay, and so distinct that it in no way can be called a real Thistle.

As the first effort of an artist, who on this occasion had to be initiated into plant-drawing, the coloured delineations now offered must be regarded as fairly creditable. Some imperfections have been amended in the full descriptions elucidative of the illustrations. Furthermore, these plates of a publication, which the Agriculture Department intends to distribute very widely over our colony, will perhaps, along with the descriptive letter-press, thus serve also educational purposes, at drawing lessons and at botanic teachings. That delineations of plants of so large a size had to be brought within such narrow pictural space, arose from the necessity of rendering this little treatise conformous to the other literary issues of the Victorian Agricultural Department.

To the Honorable the Minister and the Secretary of the Agriculture Department I beg to tender my best thanks for providing the means and for other facilitations afforded to elaborate this publication.

I.

CARDUUS LANCEOLATUS, Linné.

(Cnicus lanceolatus, G. F. Hoffmann; Cirsium lanceolatum, Scopoli.)

The Spear-Thistle, called also but inaccurately the Scotch Thistle; one of the Plume-Thistles.

Indigenous to Europe, Western Asia, and Northern Africa; here the commonest and most readily spreading of all Thistles.

A formidable plant. Height to several or even many feet, but only once flowering from the same root under ordinary circumstances. Stems and branches robust, furrowed, beset with mostly crisped hairlets. Leaves decurrent, the largest when well developed to one foot long, mostly pinnatifid, but some particularly the upper less divided, often also irregularly prickly denticulated; on the upper side rough from minute straight rigidulous hairlets; on the lower side more or less bearing a lax somewhat webby or cottony but not close vestiture, or rarely quite whitish-lanuginous; the lobes often almost semilanceolar, undivided or some incised, always terminated by a conspicuous pungent spinule. Headlets of flowers erect, comparatively large, approached by diminutive floral leaves, singly terminating branches, or two or more rather near together by the shortness of approaching branchlets. Involucre almost globular or somewhat ovate, soon open, consisting of numerous conspicuous rigid spinescent and mostly spreading bracts lanceolate or subulate-linear in form, and usually bearing some weblike vestiture below, the innermost hardly pungent. Receptacle copiously beset between the flowers with setular-capillulary bracts. Corolla of the flowers upwards purplish (shown in our analytic plate hardly of a sufficiently bright tinge), thinly tubular to the middle, towards the summit dilated and produced into five narrow lobes. Filaments of the stamens disconnected, beset with some hairlets; the five anthers united into a tube, purplish, sagittate-linear, their pollen pale. Style capillular-filiform, glabrous; the two stigmas very slender, purplish, soon exserted, coherent except at the summit. Fruits seed-like, obliquely cuneate-ellipsoid, compressed, but also latterly angular, glabrous, outside pale yet shining, smooth. Pappus somewhat shorter than the corolla, consisting of whitish hygroscopic capillulary and distantly plumous setules, finally much expanding and seceding, but often carrying the fruit with its solitary seed widely through the air, as in the case with all true Carduus Thistles. In cold climes this, like many other Thistles, is regarded as biennial; and here also it may pass into a second year if the growth becomes retarded or interrupted.

Although the specific name is constantly ascribed to Linné, it occurs already in Ray's works, if not before in Gerarde's, with precisely the same wording, and it was indicated even by J. and C. Bauhin.

Explanation of Plate I.

a. A flowering branch with its leaves, natural size.
b. 1. Longitudinal section of a headlet of flowers.
2. A complete flower, with young fruit, also bracteal bristlets and pappus.
3. A separate corolla.
4. Stamens flattened out.
5. Style and stigmas.
6. Fruit.
7. Transverse section of a fruit, showing the two cotyledons of the seed.
8. Embryo.
9. A bracteal bristlet.
10. A pappus-bristlet.
1–10. Magnified, 1 slightly so, the rest to various extent.

II.

CARDUUS ARVENSIS, Tabernæmontanus.

(Cnicus arvensis, G. F. Hoffmann; Cirsium arvense, Scopoli; Serratula arvensis, Linné.)

The perennial Thistle (called also, but incorrectly, the Canadian and the Californian Thistle), one of the Plume-Thistles, easily recognised by its perennial root and by having staminate and pistillate flowers perfected only on distinct plants, and these not always intermixed. Indigenous to Europe, Western Asia, and Northern Africa. The most mischievous of all immigrated Thistles, on account of the very great difficulty encountered in its eradication.

Height usually up to some few feet. Root thick, spreadingly penetrating to a gradually great length and depth through the soil, hence much ramified, very tenacious of vitality, the ramifications of the root brittle. Stem furrowed, variously branched, particularly the upper portion, as well as the peduncles often somewhat lanuginous. Leaves, unless the lower, sessile, usually not decurrent or only slightly so, pinnatilobed or only short-sinuated or some almost entire, often crisped, pungent-pointed and spinular-denticulated, often nearly glabrous, or on below scantily webby-lanuginous. Headlets of flowers generally stalked and somewhat paniculated, upright, of two forms; those with perfectly polliniferous anthers on individual plants distinct from those producing fertile seeds and larger, with more exserted flowers; male involucre more semi-globular; female involucre more truncate-ovate; involucrating bracts comparatively small, much appressed, from broad to narrow lanceolar and slightly fringed, the lower the broadest and short-spinulous at their apex, the upper gradually the longest and hardly or not at all pungent, all imperfectly fringed with minute hairlets, often of a rather dark hue. Receptacle bearing capillulary-setular bracts between the flowers. Corolla pale-purplish or more lilac, seldom white, its five lobes much longer than broad, bluntish, suddenly emerging from the very slender tube. Stamens alternate to the corolla-lobes, their filaments disconnected, rough; anthers arrow-shaped linear, in the male plant perfect and therefore pollen-bearing. Style thinly filiform; the two stigmas narrow but conspicuously broader than the style, and except at or towards their summit coherent, more divergent at the upper end in the female plant; fruits seed-like, truncate-ellipsoid, smooth, shining, less readily maturing than in most other species (the lithographic colouration incorrect). Pappus fragile, consisting of distantly plumous setules, finally seceding, those at least of the female plant conspicuously overreaching the corolla.

The most difficult of all our Thistles to subdue, though less copiously producing seeds fit to germinate.

Explanation of Plate II.

a. Perfect staminate plant; flowering branch with its leaves, natural size.

1. Longitudinal section of a headlet of flowers; aside an enlarged floral bract.
2. A separate corolla, the stigmas emerging.
3. Corolla, laid open, showing also the stamens.
4. A perfect anther, with part of the filament.
5. Side and back view of sterile stigmas with their style.
6. A sterile fruit.
7. Pappus.
8. A separate pappus-bristlet.

2–8. Magnified, but to various extent.

b. Perfect-pistillate plant, flowering branch with its leaves, also portion of the root and some young shoots; natural size.

1. Longitudinal section of a headlet of flowers.
2. A separate corolla, anthers and stigmas also visible.
3. A corolla, laid open.
4. A sterile anther with portion of filament.

5 and 6. Fertile stigmas with style.

7. Pappus.
8. A separated pappus-bristlet.
9. Fertile fruit.

2, 3, 4, 5, 6 and 8. Much magnified, but to various extents.

III.

CARDUUS PYCNOCEPHALUS, Jacquin.

The Shore-Thistle.

Indigenous to Middle and Southern Europe, Northern Africa, and South-Western Asia.

The specific name is derived from the headlets of flowers, generally crowded, mostly without stalks at the summit of the branches.

Height of the plant up to some few or rarely several feet; only once flowering from the same root under ordinary circumstances. Leaves much decurrent, except the lowest, when most developed to several inches long, of moderate width, frequently pinnatifid, irregularly spinular-denticulated, above seldom quite glabrous, beneath often from short hairlets webby-tomentellous, the lobes partly or irregularly incised, always ending in a short spinule. Headlets rather small and remarkably narrow, erect or diverging, usually from two to five close to each other at the end of the uppermost branchlets, without any near approach of conspicuous eaves. Involucre at flowering time almost hemiellipsoid or even, somewhat cylindric, occasionally bearing a lax webby vestiture, but not unfrequently almost glabrous, the constituting bracts rather small and smooth, pale-greenish outside, inside colourless and very shining, mostly linear-semilanceolar, but the lowest verging into an ovate form, hardly any of them strongly pungent, the upper somewhat spreading—(drawn on the whole too small, and not sufficiently pointed in the lithographic plate). Flowers much less numerous within each involucre than in most other species, linear-setular smooth bracts copiously betwixt them; corolla purplish or verging into violet colour, its lower half thinly tubular, the upper half divided to beyond the middle into five very narrow lobes. Stamens alternate to the corolla-lobes, their filaments upwards disconnected, and there beset with minute hairlets. Anthers pale, united into a tube, sagittate-linear. Style filiform, the two stigmas violet-coloured, very narrow, coherent except at the summit, encircled at the base by minute papillules. Fruit compressed, almost ellipsoid, but more attenuated towards the base; when ripe brownish or remaining pale outside and very minutely foveolar-dotted, slightly constricted below the depressed summit. Pappus somewhat or hardly shorter than the corolla, finally deciduous, the constituting capillulary setules merely subtle-ciliolated; thus this species not being a Plume-Thistle. When the branches are more broadly dilated by the decurrent leaves, then this plant as a variety bears the name Carduus tenuiflorus (Curtis). To this Thistlė is similar Carduus crispis, L., which has not yet any particular hold on Victorian soil, but is likely to invade hereafter our colony also, it being of still wider range in Europe, Asia and Africa than Carduus pycnocephalus; it can already be recognised by leaves generally less lobed, less pungent and less rigid, by less crowded headlets with involucres much more widely spreading out, by narrower involucral bracts, also more subtle-furrowed fruits.

Explanation of Plate III.

Flowering branch with its leaves, natural size.

1. Longitudinal section of a headlet of flowers.
2. A floral bract.
3. A complete flower with young fruit; also bracteal bristlets and pappus.
4. A separate corolla, the stigmas also visible.
5. A corolla, laid open, showing the stamens.
6. Style and stigmas.
7. A fruit.
8. Transverse section of a fruit; also showing the cotyledons of the seed; longitudinal section of fruit (the cotyledonar division drawn too deeply).
9. Pappus-bristlet.

1–9. Magnified, 1 slightly so, the rest to various extent.

IV.

CARDUUS MARIANUS, Linné.

(Silybum Marianum, Gærtner.)

The Spotted Thistle, called also the Maria-Thistle and the Milk-Thistle. Indigenous to Southern Europe, Northern Africa, and South-Western Asia. The generic name Silybum is recorded as the Roman one by Plinius, and originated from the vernacular of the plant in the ancient Greek language. Next to Carduus lanceolatus, hitherto the most frequent Thistle here.

Height to several feet. Flowering under ordinary circumstances only once from the same root. Stem robust, much streaked and rather angular, occasionally the upper portion somewhat beset with crisp or floccous hairlets. Leaves spacious, some to 2 feet long, the lower somewhat pinnatilobed or only sinuated, the upper with bilobed base amply clasping but otherwise often lobeless, indented, all spinular-denticulated, glabrous, shining-green and usually with whitish spots and whitish costular and venular markings. Headlets very large, soon erect, as broad as long or broader, almost flat at the base, singly terminal, often conspicuously stalked. Involucral bracts mostly of foliaceous texture, the outermost broadly stipitate, nearly cordate or semiorbicular, spinular-denticulated and spinescently long-acuminated; the more middle bracts very long and much spreading, towards the base entire and almost stalklike, though also broad and flat, thence passing into a deltoid or cordate-orbicular large dilatation, concave or channelled, irregularly spinular-denticulate and gradually terminating in a long spinule; the innermost bracts almost linear-lanceolar, more or quite entire and somewhat scarious. Flowers numerous, with very copious setulous-capillulary persistent bracts between them. Corolla upwards purplish, reddish or rarely white to beyond the middle slender-filiform, then suddenly widening and cleft into five somewhat unequal linear-elliptic lobes. Stamens alternate to the corolla-lobes, their filaments for the greater part of their length connate into a tube; anthers purplish, linear-sagittate. Style filiform; stigmas somewhat broader, coherent except at the summit, purplish, soon exserted. Fruit ovate-ellipsoid, truncate, considerably compressed, smooth, with a protruding apex, outside dark-brown, with slight paler variegation. Bristlets of the pappus flattish, ciliolar-serrulated or those of the outer flowers slightly plumous, connectedly deciduous, not very much longer than the ripe fruit.

Explanation of Plate IV.

Flowering branch with its leaves, natural size.

1. Longitudinal section of a headlet of flowers.
2. A complete flower with young fruit and pappus.
3. Stamens with summit of the corolla and only upper portion of the filamental tube.
4. Stigma and upper portion of the style.
5. A bracteal bristlet and a pappus-bristlet.
6. A fruit.
7. Transverse section of a fruit, showing also the cotyledons of the seed.
8. Embryo.

2–8. Enlarged, but to various extent.

V.

ONOPORDON ACANTHIUM, Linné.

The Scotch Heraldic Thistle.

The generic name was written by Linné Onopordum, but the writing generally adopted is in accordance with the name as given by Plinius from ancient Greek authors.

Indigenous to Europe, Western Asia, and Northern Africa. Here less obstructive than the other Thistles. Height up to some few or rarely many feet, but often dwarfed. Flowering under ordinary circumstances only once from the same root.

All over or particularly the leaves beneath more or less webby-lanuginous. Leaves conspicuously decurrent, except those at the root, in outline from rhomboid-ovate to narrow-lanceolar, short-lobed or only indented, always irregularly spinescent-denticulate, sometimes glabrescent above, the lower to one foot long and occasionally some deeply cleft. Headlets of flowers singly terminating branches or approximated branchlets. Involucre truncate-globular, consisting of numerous rather small spreading upwards linear-subulate very pungent bracts, the lowest bent downwards. Receptacle honeycombed and denticular-foveolate, but devoid of setular-capillary bracts. Corollas upwards purplish-red or rarely quite white, the lower half thinly filiform, the upper half suddenly turgid, somewhat irregularly cleft into five narrow bluntish lobes. Stamens alternate to the corolla-lobes, their filaments almost glabrous ; anthers connate, linear-sagittate, purplish. Style capillulary; stigmas narrowly semicylindric, coherent except at the summit, with a slight basal enlargement. Fruits somewhat compressed, nearly ovate, longitudinally angular-lined, outside brown or greyish and occasionally black-spotted, conspicuously wrinkled, supported within its receptacular hollow by only a minute bract. Pappus-bristlets numerous, subtly ciliolar denticulated, at the base connate, sometimes reddish.

Explanation of Plate V.

Flowering branch with its leaves, natural size (lithographic colouration green instead of greyish).

1. Longitudinal section of a headlet of flowers.
2. Alveolar bracts of receptacle, one yet retaining its fruit.
3. A complete flower, with young fruit and pappus.
4. A corolla laid open, showing also stamens, style and stigmas.
5. A stamen separated.
6. Style and stigmas.
7. Side view of a fruit.
8. Fruit, presenting its summit.
9. Longitudinal section of a fruit across the cotyledons.
10. Longitudinal section of a fruit parallel to the cotyledons.
11. Transverse section of a fruit.

1–10. Enlarged, but to various extent.

VI.

CENTAUREA CALCITRAPA, Linné.

The Star-Thistle.

The specific name has been chosen in allusion to the very spinular flower-headlets, its etymology being derived from the Latin name of an ancient contrivance to impede the progress of horsemen in warfare. The British vernacular has been in use since medieval times. Indigenous from Middle Europe to Western Asia and Northern Africa.

Flowering under ordinary circumstances only once from the same root. Neither tall nor robust, often much spreading and very branched, imperfectly beset with short hairlets, only occasionally reaching a height of 3 feet. Leaves rather small, unless those near the root, always lax, simply sessible, pinnatilobed or the upper gradually undivided, the basal leaves sometimes doubly dissected, the lobes and undivided portion always narrow and distantly or imperfectly denticulated. Headlets of flowers singly terminal and lateral, almost sessile. Involucral bracts glabrous, at their lower portions closely appressed, dilated, pale, uniting into an almost conic-ovate form, the upper ending in a spreading comparatively long and strong spinule, excepting the innermost bracts; the spinule towards its base denticulate-spinular and somewhat channelled. Receptacle beset with numerous capillulary-setular white bracts. Flowers of nearly the same length, not very numerous, the outermost generally sterile. Corolla purplish in the majority of the plants, its tube very thin, widened towards the summit, the five lobes narrow; corolla of the circumferential flowers longer lobed. Stamens irritable, alternate to the corolla-lobes, their filaments beset with very minute papillular hairlets; anthers reddish, connate, linear-sagittate. Style capillulary, bearing papillular hairlets at the upper end; stigmas narrowly semicylindric, coherent except at the top. Ripe fruit glabrous, cuneate-ellipsoid, somewhat compressed, slightly biangular, at the base unilaterally impressed or almost excised, outside pale with darker striolate spots (the lithographic colouration on the plate incorrect). Pappus absent.

A very closely allied plant, showing a transit to Centaurea Iberica (Stephan), differs chiefly in having a short pappus, at all events to the fruits of the marginal flowers. A variety, here also already immigrated, in which the lower involucral bracts are terminated only in very short spinules, approaches Centaurea Pamphylica (Boissier and Heldreich).

Explanation of Plate VI.

Flowering branch with its leaves, natural size.

1. Longitudinal section of a headlet of flowers, slightly enlarged.
2. A separate corolla, showing also style and stigmas.

3. A corolla laid open, showing thus the stamens.
4. A stamen detached.
5. Style and stigmas.
6. A fruit.
7. Transverse section of a fruit, exhibiting the cotyledons.
8. Embryo.
2–8. Enlarged, but to various extent.

VII.

CENTAUREA MELITENSIS, Linné.

The Malta-Thistle.

Indigenous to Southern Europe, Northern Africa, and South-Western Asia.

Neither this nor any other Centaurea is so formidable as any of the true Carduus-Thistles, nor do they so easily spread from seeds to far distances; flowering under ordinary circumstances only once from the same root. Vestiture short, soft, grey, partially evanescent. Height of plant to three feet or exceptionally more. Stem erect, usually few-branched, foliaceously dilated at its angles. Radical leaves of good size, narrowed into a conspicuous petiole, pinnatifid, the end-lobe largest; cauline leaves rather small, long decurrent beyond the point of affixion, from narrow-lanceolar to broad-linear, quite entire or somewhat denticulated, the floral leaves much shortened. Headlets of flowers absolutely terminal, or some at the ends of short but seldom crowded branchlets almost axillary, comparatively small. Involucre turgid, but generally less broad than given in the illustrative plate, also less spreading and spinulous in the downward portion, often participating in the general vestiture, its constituent bracts downward much appressed, near the upper end most of them spreadingly spinular-denticulated, and except the lowest and innermost ending into a spinule of a length less than that of the whole bract. Receptacle beset with numerous capillulary-setular bracts. Corollas yellow; those of the peripheral flowers somewhat enlarged and generally sterile; tube thinly filiform, rather suddenly widening, the dilated portion nearly as long as the five narrow and acutish lobes. Stamens alternate to the corolla-lobes; their filaments glabrous; the anthers shaped like a narrow elongated arrow-head, but slightly contracted at the middle. Style capillulary, with circles of papillular hairlets at the upper end, otherwise almost glabrous. Stigmas semicylindric, somewhat curved, connate except at the summit. Fruit ellipsoid, moderately compressed, at the top truncated, above the narrow base unilaterally much impressed, scantily covered with almost imperceptible hairlets, shiningly greyish outside. Pappus-bristlets flattish, but extremely narrow, scarcely subtle-ciliolated, those of the outer row less than half the length of those of the inner series, but the latter still shorter than the capillulary bracts of the receptacle and hardly as long as the fruit.

A closely allied species, namely Centaurea solstitialis (Linné), the St. Barnaby's Thistle, which is of the same geographic range in its native countries, has also already found its way into the Colony Victoria for permanent domiciliation; it remains however as yet less frequent, though in the countries at the Mediterranean Sea it seems more common than Centaurea Melitensis, which does neither claim British natality in the way of Centaurea solstitialis. The latter may be recognised already by the terminating spinules of the involucral bracts being considerably longer than the lamina of bracts, also by the outside darker colour of the fruits, with less attenuation at the base, and by the longest pappus-bristlets about doubly exceeding the length of the fruit, irrespective of some minor characteristics.

Explanation of Plate VII.

Flowering branch with its leaves, natural size.

1. Longitudinal section of a headlet of flowers, slightly enlarged.
2. Pappus-bristlets.
3. A separate corolla, unexpanded.
4. A separate corolla, expanded.
5. A corolla laid open, showing thus the stamens.
6. An anther with part of its filament.
7. Style and stigma.
8. A fruit, its pappus shed.
9. Transverse section of a fruit.
10. Embryo separated, showing cotyledon and radicle.
11. Whole pappus.
12. Inner pappus-bristlets.

2–7. Enlarged, but to various extent.

VIII.

KENTROPHYLLUM LANATUM.

De Candolle and Duby (from Necker), Carthamus lanatus, Linné.

The Saffron-Thistle.

The generic name is derived from the spinular-prickly leaves, the vernacular from the affinity to the spurious Saffron-Plant, Carthamius tinctorius, L.

Indigenous to Southern Europe, Northern Africa, and South-Western Asia.

Flowering under ordinary circumstances only once from the same root, but it may pass into a second year of growth. Stem strongly streaked, to about three feet high, unless exceptionally taller; its upper portion as well as the branches somewhat webby-lanuginous. Leaves stiff, bright-green, scantily lanuginous or glabrescent and slightly viscidulous papillular, prominently venular, the lowest stalked and often pinnatilobed, their end lobe much the largest; the upper leaves gradually smaller, clasping, producing short narrow spinular lobes, pungent-pointed, some of the lobes reduced to denticles. Headlets of flowers large, solitary, surrounded by floral leaves. Involucral bracts spacious, the outer upwards of leafy texture and colour, to some extent webby-lanuginous, from a broad base narrow or linear lanceolar, spinular-lobed or denticulated, the innermost narrower, nearly or quite entire and somewhat scarious. Receptacle copiously beset with setular-linear smooth bracts. Flowers numerous on each receptacle and within each involucre, uniform, unless some few of the outer sterile and devoid of a pappus. Corollas intensely yellow, more deeply so in age, towards the summit marked by dark longitudinal stripes, their tube thinly filiform, suddenly widening near the five linear lobes, its base orbicularly dilated. Filaments of the stamens near the upper end barbellate; anthers connate, sagittate-linear. Style capillulary, glabrous; stigmas thinly semicylindric, except at the summit connate. Fruits seed-like, glabrous, almost semiovate, distinctly quadrangular, distinctly wrinkled, their basal attenuation somewhat oblique, outside pale or turning light-brownish. Pappus finally light-brownish, its outer scalelets very much shortened and partly truncate-bluntened, the inner gradually much longer and pointed, some exceeding twice the length of the fruit, all broadish-linear and minutely ciliolated.

Explanation of Plate VIII.

Flowering branch with its leaves, natural size.

1. Longitudinal section of a headlet of flowers, slightly enlarged.
2. A separate corolla, showing also stigmas.
3. A corolla laid open, the summit removed, the stamens also exhibited.
4. A separate anther with upper part of filament.
5. Style and stigma.
6. Side-view of a fruit with its pappus.
7. Top-view of a fruit with its pappus.
8. Scalelets of pappus.
9. Transverse section of a fruit, showing also the cotyledons.
10. Embryo, showing cotyledons and radicle.

2–10. Enlarged, but to various extent.

IX.

XANTHIUM SPINOSUM, Linné.

The Bathurst-Burr.

The generic name, if really indicated already by Dioskorides, is derived from a yellow pigment of this kind of plants, further from the yellow colour of the spinules. The specific name occurs first in Professor R. Morison's writings (1699). Assumed to be a native of South-Western Asia, though also thought to have originated in South-Western America. Flowering under ordinary circumstances only once from the same root. Up to some feet high, but generally more dwarfed. Branches finally almost glabrescent. Leaves on short stalks, never very large, in outline often rhomboid-lanceolar, usually three-lobed, otherwise almost or quite entire or seldom somewhat pinnatilobed, above dark-green and beset with appressed scattered short hairlets, more so along the axis and venules, beneath from a thin close vestiture greyish or whitish, the lobes acute, the terminal lobe much the longest; primary venules prominent. Spinules arising from near the base of the leaves, very conspicuous, acicular, usually ternate, but at the base connected, yellow, each representing the primary axis and the secondary of an undeveloped leaf. Flower-headlets quite small, almost or wholly unisexual; the staminate more terminal, globular, single or occasionally some few together, the pistillate usually in lower axils, all nearly sessile. Involucre of the staminate flowers consisting of very small mostly lanceolar puberulous bracts in several rows, almost free from each other; receptacle extremely short (drawn too long in the illustrative plate), individual flowers many, partly supported by solitary minute bracts; corolla beset with very short hairlets outside; lobes five, very short; tube almost obverse-conic; filaments of the stamens connate into a cylinder or separating; anthers disconnected, soon seceding, linear, curved-apiculate, finally spreading; pollen pale; style, stigmas and ovulary rudimentary or absent. Pistillate flower-headlet often solitary, soon bent downward, its involucre ellipsoid, consisting of two enlarging and bilocularly concrescent bracts, all over beset with numerous small very spreading hooked inwardly (not outwardly as drawn) yellowish or brownish spinules, open at the summit, including only two flowers, bearing a short close vestiture except at the spinules, at last hardening, often terminated by two straight spinules; corolla none or rudimentary. Stamens absent; style thinly cylindric; stigmas two, capillulary-semicylindric, emerging and finally spreading. Fruits seed-like, elongated, concealed, filling the cavities, somewhat compressed, one-seeded, dark-coloured outside, glabrous, devoid of any pappus.

Explanation of Plate IX.

Flowering branch with its leaves and spinules, natural size.

1. Top view of a staminate headlet.
2. Base view of a staminate headlet.
3. A staminate flower detached, with its own bract.
4. A staminate flower laid open.
5. The five stamens separate.
6. Front and back view of anthers.
7. Style and stigmas, young.
8. Involucre and receptacle of staminate headlet, the flowers removed.
9. Fruiting involucre.
10. Transverse section of fruiting involucre, showing its two cells with seed-bearing fruits.
11. An involucral spinule, separated.
12. Embryo, showing the cotyledons and radicle.

1–12. Enlarged, but to various extent.

By Authority: ROBT. S. BRAIN, Government Printer, Melbourne.

P. Ashley. Del. et Lith. F. v M. Dir. Stebbing, Addy & Co Impr

CARDUUS LANCEOLATUS (*Linné*).

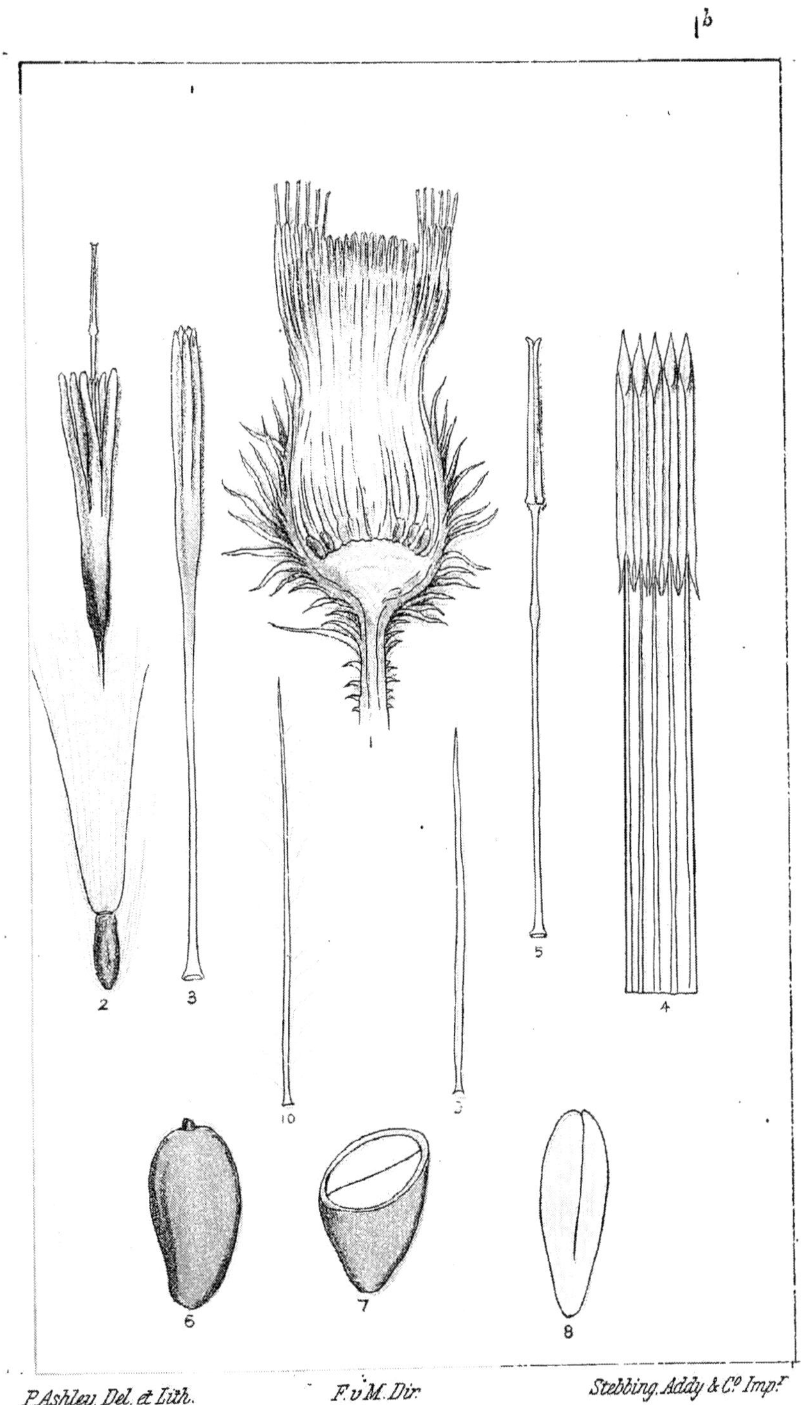

P. Ashley. Del. et Lith. F. v M. Dir. Stebbing, Addy & Co. Impr.

CARDUUS LANCEOLATUS (Linné).

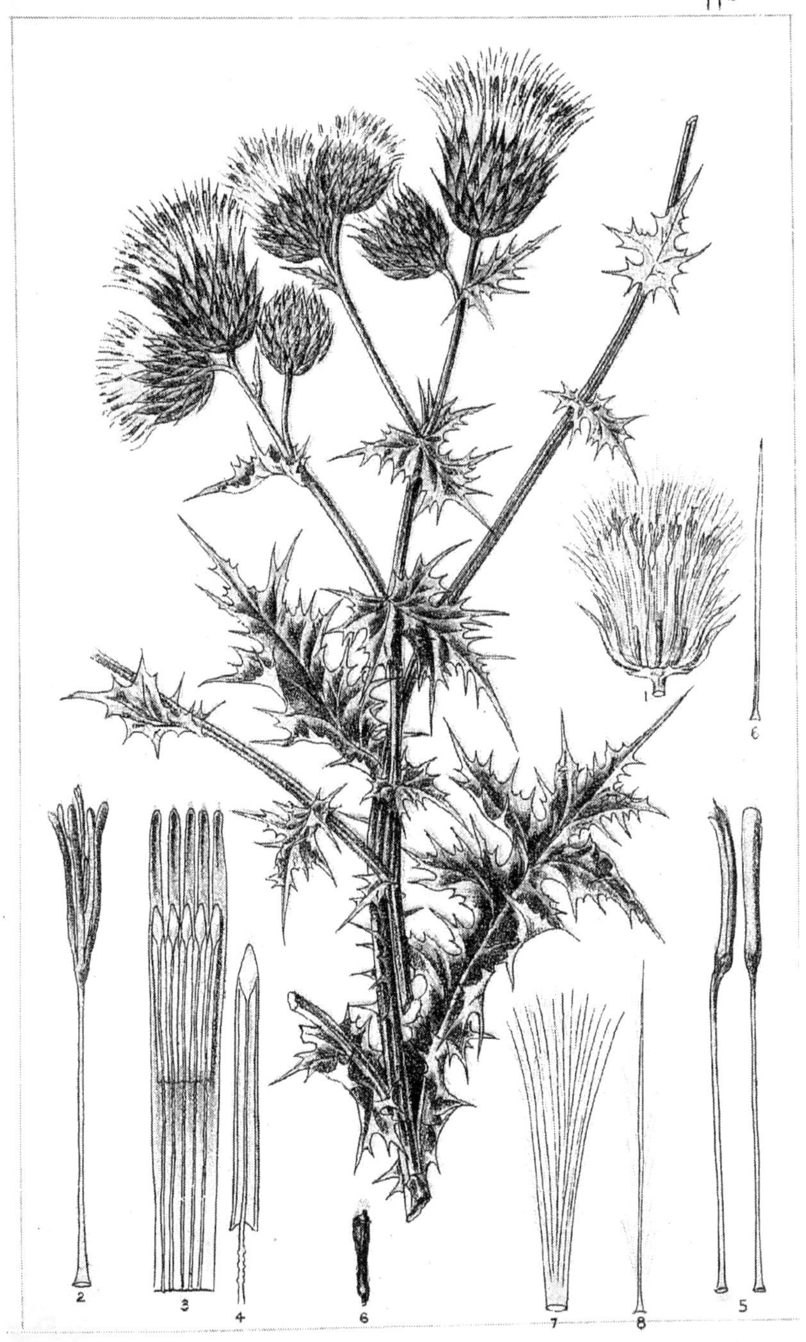

P. Ashley. Del. et Lith. F. v M. Dir. Stebbing. Addy & Co. Impr.

CARDUUS ARVENSIS (Tabernaem).

♂

P. Ashley. Del. et Lith. F. v M. Dir. Stebbing. Addy & Co. Impr.

CARDUUS ARVENSIS (*Tabernaem*).

♀

P. Ashley. Del. et Lith. F. v M. Dir. Stebbing. Addy & C° Imp^r

CARDUUS PYCNOCEPHALUS (Jacquin).

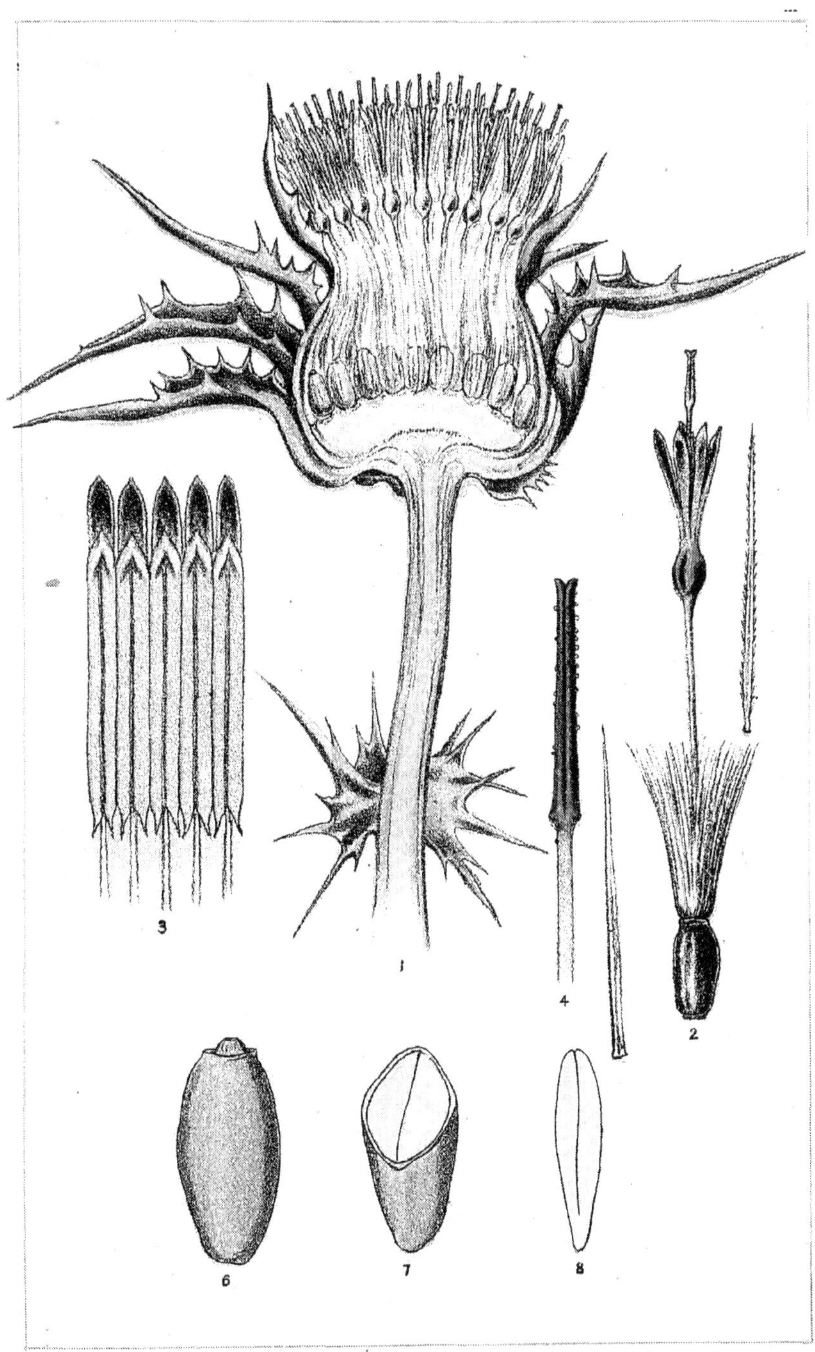

P. Ashley. Del. et Lith. F. v M. Dir. Stebbing. Addy & Co. Imp.

CARDUUS MARIANUS *(Linné)*.

IV^b

F. Ashley, Del. et Lith. F. v M. Dir. Stebbing Addy & Co. Impr.

CARDUUS MARIANUS *(Linné).*

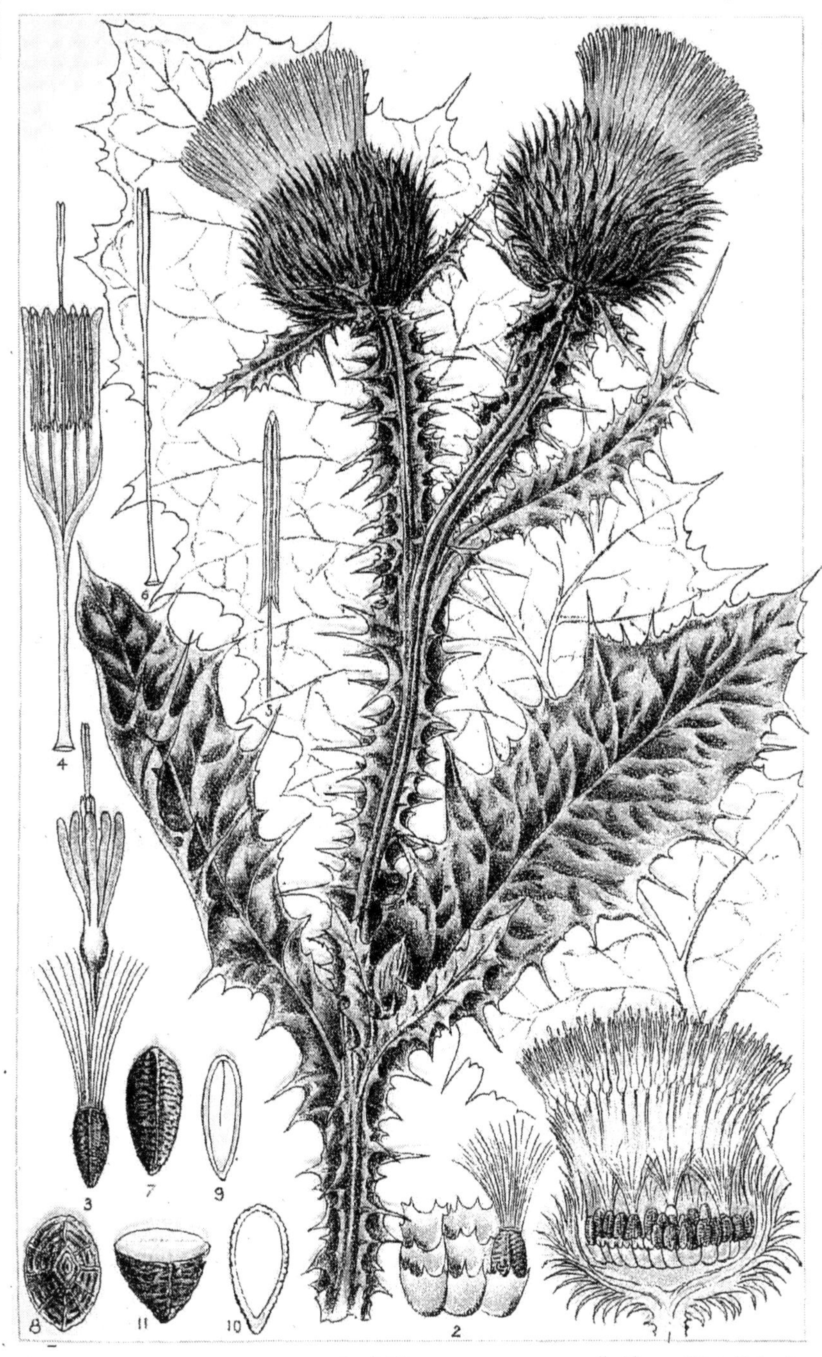

P. Ashley. Del. et Lith. *F. v M. Dir.* *Stebbing Addy & C°. Imp^r*

ONOPORDON ACANTHIUM *(Linné)*.

Ashley, Del. et Lith. F. v M. Dir. Stebbing, Addy & Co. Impt.

CENTAUREA CALCITRAPA (*Linné*).

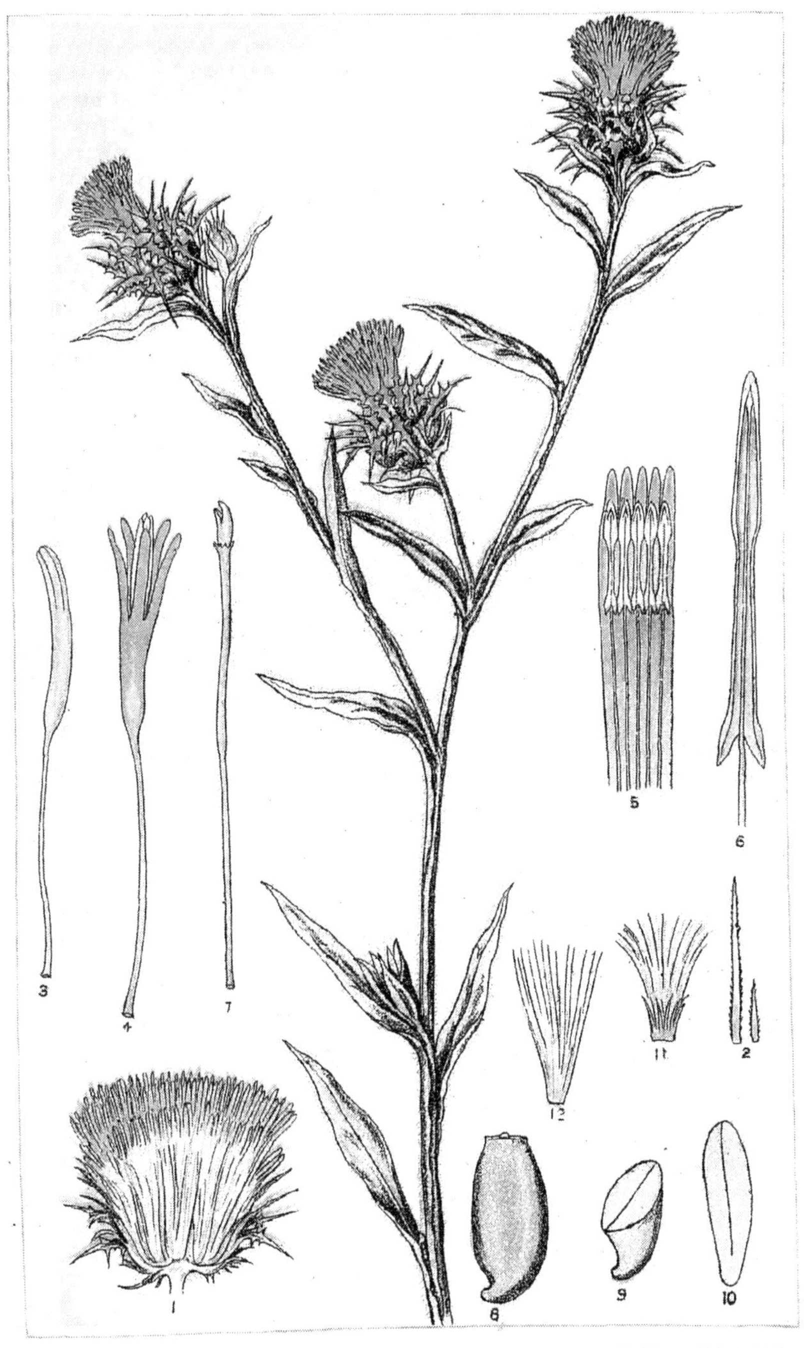

P. Ashley Del. et Lith. F. v M. Dir. Stebbing, Addy & Co Imp

CENTAUREA MELITENSIS (*Linné*).

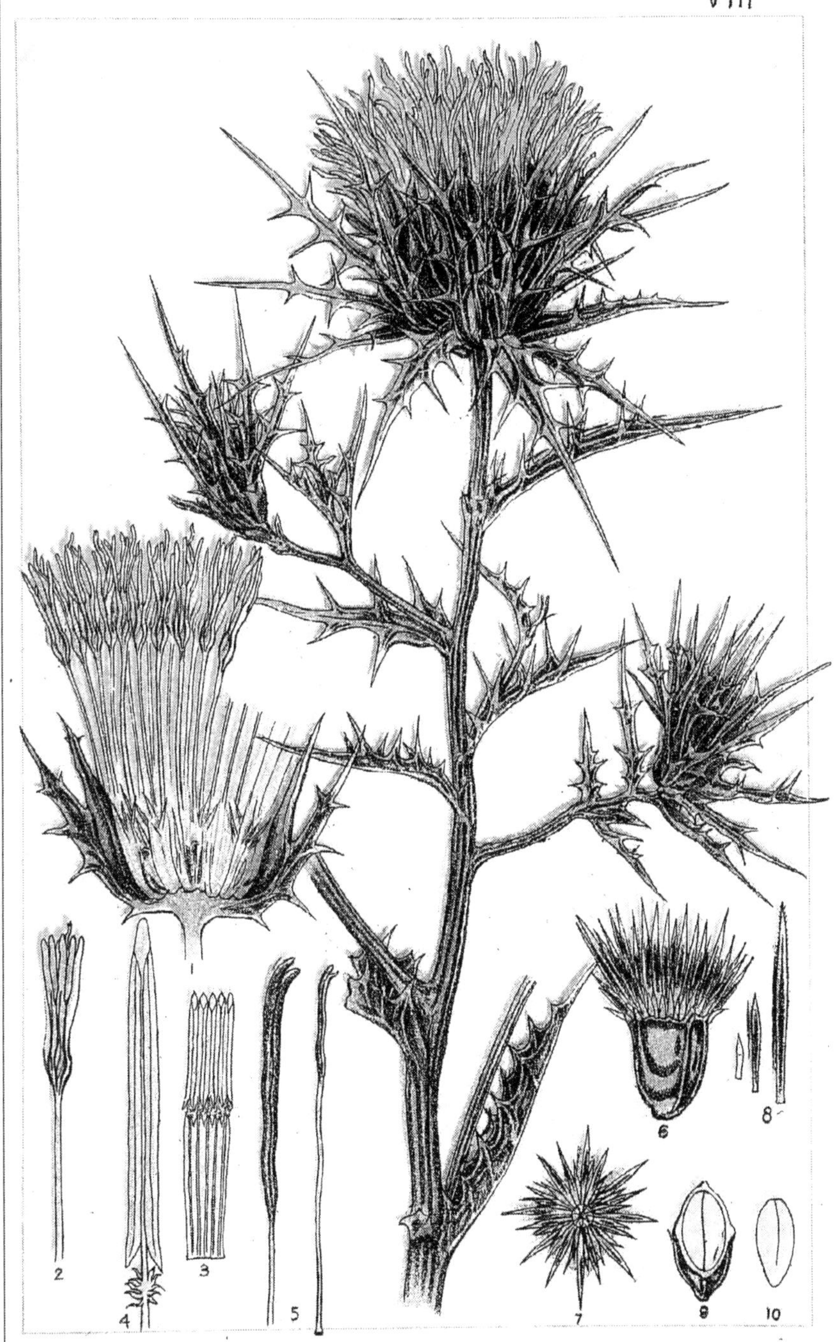

P. Ashley. Del. et Lith. F. v M. Dir. Stebbing, Addy & Co Impr.

KENTROPHYLLUM LANATUM (Necker).

P. Ashley, Del. et Lith. F. v M. Dir. Stebbing, Addy & Co. Impr.

XANTHIUM SPINOSUM (*Linné*).

Printed in the USA
CPSIA information can be obtained
at www.ICGtesting.com
LVHW020845250923
759136LV00008B/1338